Homeostasis, Nature's complex balancing act,

is our body's maintenance of its own chemistry.

The autonomic nervous system supports our lives

and homeostatic processes keep us in good health.

How many of us could
survive if our breathing,

walking or eliminating
were dependent upon thought

related processes?   To
what end are we as humans

freed from those
functions?

For the genus homo and
species homo sapiens, the

last surviving species of
our 2.5 million year old

genus, I must preclude
that we have been freed as

have **all** the breathing
species of the planet to *freely*

think.   After all, the noble
quest to understand the

origin of the cosmos and
meaning of life

may indeed only be
revealed using the control

center of life and thought,
the "mind".   Cerebration,

the ephemeral process
which creates ideas, hopes,

emotion and dreams,
suddenly present and as

quickly gone. Based upon
Nature's proclivity for the

variety and complexity of
most of the

breathing entities it
creates and the 14 billion years
it

has expended to date
creating them, one

might plausibly reason
that it

just may be that "thinking"
itself was one of the more

favorable outcomes of its efforts. Or not.

Human beings certainly have evolved into if not the

most *dependably* wise thinkers, certainly the most

capable of showing promise toward that

end. " 'I think therefore I am'," may best

be stated, *"I think therefore I may become."*

Become whom?   That is
an evolutionary question

indeed, and is it possible
that Nature has truly

unchained the succession
of life on the planet from

the labor of maintenance,
for the glory of thought?

With our accumulated
history, increasing knowledge

and potential for elevated wisdom, that

could mean growing more

aware of the moment and becoming more capable

as a species of, critical and insightful thinking

toward a peaceful and harmonic solution to life on

this planet.   Possibly so, or for that matter if we are

that stupid at the least or
that unfortunate at best

we may "think" our
collective beings into

armageddon and oblivion.

For within the ever
increasing cerebral capacity

of the genus homo lay the
seeds of their

enslavement as well as their freedom. For just as

wet clay is molded on the potter's wheel and

hardened in the kiln, we are beset by mind forged

manacles. Before we realize it, the potential for self

realization is stifled as the spontaneity of a child's

expressions and thoughts may soon harden into

mind dulling habit.

Creativity is often debilitated due to early educational

and emotional experiences and only through

focusing the lens of introspection, might we ever

recapture and develop our being-potential as we

grow in age.

There are hundreds if not thousands of

psychotherapies, awareness seminars and

medications all attempting

to pull the mind out of the rut it may find itself in.

This impasse is the habitual reflexive type of thought

(systemic thought pattern)
which perpetuates events

that we have passed
through in time, but not in
mind.

It is this rut which prevents
us from focusing 100% of

our power of thought to
maximally embrace the

"present", as Nature has *most certainly* provided for.

We learn through repetition until recall is almost

instantaneous as we retrieve our memories from the

neurons through which they were originally thought.

Our thought involves association and by linking one

concept to another, vocabulary, language and

intelligence are
constructed.   A problem arises
due

to the fact that just as
physical pain lingers beyond

that of physical pleasure,
an exact parallel exists

with regard to psychical
pain and pleasure. Our

thought patterns develop in the normal process of

thinking.

The more we *feel* a thought experience as

hurtful and damaging, the greater are our futile

attempts to relegate it to our unconscious and

the more likely its inveterate fixation in our mind.

These thoughts and associated mechanisms

should be more thoroughly understood so that we

might ameliorate their debilitating effect.

The cerebral cortex or forebrain consists of the "old

mammalian brain" (limbic system) and cerebrum

(neocortex). All
mechanisms necessary for
thought

and emotion lie therein.
Evolving through time the

neocortex has
mushroomed in size
surrounding the

limbic system, the latter
being the limit of many

primates' evolutionary
statement to date. The limbic

system is known as the
"emotional" control center in

all mammals. Almost 100
years ago patients with

rabies exhibited anguish,
rage, terror, etc. provoked

by the virus' infiltration into
the hippocampus, a

major part of the limbic
lobe; medical science has

subsequently confirmed this many times over. Our

interconnecting neurons, "wires" if you will, conduct

all external and internal impulses including

thought. How we are wired and which

electrochemical impulses are transmitted, are the

determinants of our specific reactions. We condition

ourselves until our thought patterns become

repetitive and rote through habitual repetitive

thought.

Our *habitual thought patterns* develop into reflex-like

actions as the conducting power of the same nerves

used for the original
thought/s increase with the

frequency of their recall
from memory.   Our basic

or unique DNA determined
nature

(psyche/personality) and
the experiences of our

early vulnerable years to
some extent, will

determine these recurring, reflexive-type thought

reactions throughout our lives.

Our emotions do not depend upon simple reflex

mechanisms, as the hypothalamus is involved with

higher limbic centers in this processing. The

precision of the neuron groups involved in our

emotions are however "comparable" to the reflexive

behavior needed for our survival throughout

evolution controlled by the autonomic nervous

system and "largely" unconcious. In the strictest

sense, neurotic behavior is habitual and not reflexive

because the stimulus originates cerebrally with no

peripheral nerve excitation, your behavior being

reflex-like with the perception of your particular

psychical or physical stimuli.

A psychical blueprint
begins to emerge, consisting of

a genetically founded and
directed psychic system

more complex than our
physical system. Our

nature-psyche-personality,
call it what you will, is

determined at conception
and minimally altered by

nurture or circumstance.

A psyche that presents as artistic, caustic,

aggressive, romantic, hedonistic, controlling,

passive, etc. is blueprinted from birth. The

person will react predictably to a given stimulus and

it may therefore be considered a triggering stimulus

for their particular psyche.
Ultimately the response

is neurotic behavior.

Photography offers us a
way to clarify the

composition of this
psychical blueprint and

associated processes
explained in detail on pages

16-19.   The photographer's pre-treated paper

seemingly blank is like the *psychical blueprint*; the

developing solutions are like the *triggering stimuli*

interacting with the pre-treated paper (psychical

blueprint); the developed print is like the

response , variant and dependent upon the type of

solution (triggering stimulus) affecting the paper.

The response is always a reflexive-like behavior.

PSYCHICAL BLUEPRINT:

Our structural nuclear DNA which reacts in part to

environmental circumstance through our organic,

electro-chemical beings. This blueprint determines

our psychical and physical traits. Based upon our

psychical traits we extend our beings into society

with mental predispositions, a.k.a. "*primary psychical*

need/s". *These needs will influence our behavior for*

life, and occur in <u>differing</u>
<u>*type, number and*</u>

<u>*calibration*</u> *within every*
human being on Earth.   The

combination and
domination of one or more of
an

individual's primary
psychical needs and the extent

to which an individual can
sate them will be the

determinant of their attitude and social behavior. In

identifying specific *primary* psychical needs this

writing will forgo broad categorization of *basic*

human needs such as

love, order, meaning, purpose, self-respect and

other's respect, as they are *"need universals"* of the

normal human psyche.

Some primary psychical needs are the *need* for:

independence, security, acceptance, risk-adventure,

romance, attention, knowledge, power, solitude,

control, social interaction, achievement, beauty

(an aesthete), etc.

TRIGGERING STIMULI:

Frustration, anger, rage, fear, sorrow, hostility,

disgust, etc., any one or more of which may be

triggered by the lack of the fulfillment of primary

psychical need/s.

RESPONSE:

The habitual reflexive type thought/s (*systemic*

patterns of thought) initiated by any triggering

stimulus. This response is essentially reflexive as

the thought stream requires no associative cognition

other than the feeling of a triggering stimulus.

Neurotic behavior will follow in the form of

depression, withdrawal, phobia, compulsion,

obsession, anxiety/panic attacks, guilt etc.

Our basic physical needs are far less subtle than our

primary psychical needs, in that the deprivation of

our physical needs can lead to a permanent

shutdown of the entire system, as in death. The lack

of fulfillment of our more subtle primary psychical

needs will cause one to misdirect their energy by

fixating on the unfulfilled need. Concentration and

productive areas of
endeavor falter as

the need is magnified out
of proportion due to the

focus and dwelling upon it.
Triggering stimuli

caused by unfulfilled
primary psychical need, cause

systemic patterns of
thought and in turn defenses

and neuroses. The need
must either be satisfied or

otherwise eliminated for
peace of mind to ensue, lest

the energy which failed to
sate the need, divert ad

nauseum to fuel the stimuli
and neuroses that follow.

Can we break the cycle of
unfulfilled need, triggering

stimulus and response?
Are the emotions and

behavior habituated for so long, at all tractable?   If

one is unaware of their psychical needs there is little

if any chance at all; even given your knowledge of

one or more of your psychical needs, one still

may not break the cycle were the need to remain

unsatisfied.   We are our
own worst

enemies and conversely
can be our own best

friends, providing we use
our self knowledge with

unabated objectivity.   We
are the ones who should

be aware of our emotional
history better than

anyone. Yet most people
are too busy handling the

world from the outside in
to take the time, have the

knowledge, wherewithal,
etc. to make the tedious

effort of handling the world
from the insight out

(a.k.a. introspection); enter
the epidemic of neuroses

and drugs.

The following two
examples will illustrate familiar

labels for pandemic
habitual neurotic reflexive

patterns of thought:
"*Phobia*" is what is called the

projection of a triggering
stimulus (fear) onto an

object thereby
externalizing it so that it may
be

controlled or avoided;
were the fear to be projected

onto an action we label it
"compulsion."

Regardless, the phobia or
compulsion will always

remain until the entire
systemic pattern of thought

causing the behavior is
dismantled by exposing the

cause of the stimulus.
"Depression" often deals with

past pain linked to a
present, by systemic thought

patterns, which is totally
different from that past. A

person who was obese and
now slender who cannot

"let go" of their
"overweight" persona even
though

years have passed, is but
one example.

The responsibility for the
equilibrium of mind and the

spontaneity of thought that
follows from it, which

Nature has surely
provided for, must inevitably be

our own. There are no
magic bullets nor instant

cures, and the hundreds of psychotherapies and

self-help seminars will rarely eliminate the problem.

Once you have the insight that *your* triggering

stimulus is the precursor of *your* habitual reflexive

thought pattern/s, you have the opportunity to

consciously break that circuit instead of habitually

replaying it and suffering its resultant behavior. You

may begin to live in the current moment ready for

active insightful thinking rather than mired in reactive

habitual thought.

The determination must be made that life is a

unidirectional linear event, and any effort to

change yesterday is futile. What *is* possible

is the mindful insight and determination we bring to

our present to ameliorate the deleterious effects that

our past has imprinted upon our psyche.

REFERENCES

Darwin, Charles, The Expression Of The
Emotions In Man And Animals,
University of Chicago, Chicago, Ill., 1965.

Changeux, Jean-Pierre, Neuronal Man, Pierre-Changeux, 1985.

www.ingramcontent.com/pod-product-compliance
Lightning Source LLC
Chambersburg PA
CBHW021928170526
45157CB00005B/2226